Standard Grade | Credit

Mathematics

Leckie × Leckie

First exam published in 2001.
Published by Leckie & Leckie, 8 Whitehill Terrace, St. Andrews, Scotland KY16 8RN tel: 01334 475656 fax: 01334 477392
enquiries@leckieandleckie.co.uk www.leckieandleckie.co.uk

ISBN 1-84372-308-5

A CIP Catalogue record for this book is available from the British Library.

Printed in Scotland by Scotprint.

Leckie & Leckie is a division of Granada Learning Limited, part of ITV plc.

Acknowledgements

Leckie & Leckie is grateful to the copyright holders, as credited at the back of the book, for permission to use their material.
Every effort has been made to trace the copyright holders and to obtain their permission for the use of copyright material.
Leckie & Leckie will gladly receive information enabling them to rectify any error or omission in subsequent editions.

[BLANK PAGE]

C

2500/405

NATIONAL
QUALIFICATIONS
2001

WEDNESDAY, 16 MAY
1.30 PM – 2.25 PM

MATHEMATICS
STANDARD GRADE
Credit Level
Paper 1
(Non-calculator)

1 **You may NOT use a calculator**.

2 Answer as many questions as you can.

3 Full credit will be given only where the solution contains appropriate working.

4 Square-ruled paper is provided.

FORMULAE LIST

The roots of $ax^2 + bx + c = 0$ are $x = \dfrac{-b \pm \sqrt{(b^2 - 4ac)}}{2a}$

Sine rule: $\dfrac{a}{\sin A} = \dfrac{b}{\sin B} = \dfrac{c}{\sin C}$

Cosine rule: $a^2 = b^2 + c^2 - 2bc \cos A$ or $\cos A = \dfrac{b^2 + c^2 - a^2}{2bc}$

Area of a triangle: Area $= \frac{1}{2}ab \sin C$

Volume of a cylinder: Volume $= \pi r^2 h$

Standard deviation: $s = \sqrt{\dfrac{\sum(x - \bar{x})^2}{n-1}} = \sqrt{\dfrac{\sum x^2 - (\sum x)^2 / n}{n-1}}$, where n is the sample size.

KU	RE

1. Evaluate

$$3 \cdot 1 + 2 \cdot 6 \times 4.$$

2

2. Evaluate

$$3\tfrac{5}{8} + 4\tfrac{2}{3}.$$

2

3. Given that $f(m) = m^2 - 3m$, evaluate $f(-5)$.

2

4. Solve **algebraically** the equation

$$2x - \frac{(3x - 1)}{4} = 4.$$

3

5. A furniture maker investigates the delivery times, in days, of two local wood companies and obtains the following data.

Company	*Minimum*	*Maximum*	*Lower Quartile*	*Median*	*Upper Quartile*
Timberplan	16	56	34	38	45
Allwoods	18	53	22	36	49

(*a*) Draw an appropriate statistical diagram to illustrate these two sets of data.

3

(*b*) Given that consistency of delivery is the most important factor, which company should the furniture maker use? Give a reason for your answer.

1

[Turn over

KU	RE

6. A is the point (a^2, a).

T is the point (t^2, t), $a \neq t$

Find the gradient of the line AT.

Give your answer in its simplest form. **3**

7. A garage carried out a survey on 600 cars.

The results are shown in the table below.

		Engine size (cc)			
		0–1000	1001–1500	1501–2000	2001+
Age	Less than 3 years	50	80	160	20
	3 years or more	60	100	120	10

(*a*) What is the probability that a car, chosen at random, is less than 3 years old? **1**

(*b*) In a sample of 4200 cars, how many would be expected to have an engine size greater than 2000cc **and** be 3 or more years old? **2**

KU	RE

8. The diagram below shows part of the graph of $y = 4x^2 + 4x - 3$.

The graph cuts the y-axis at A and the x-axis at B and C.

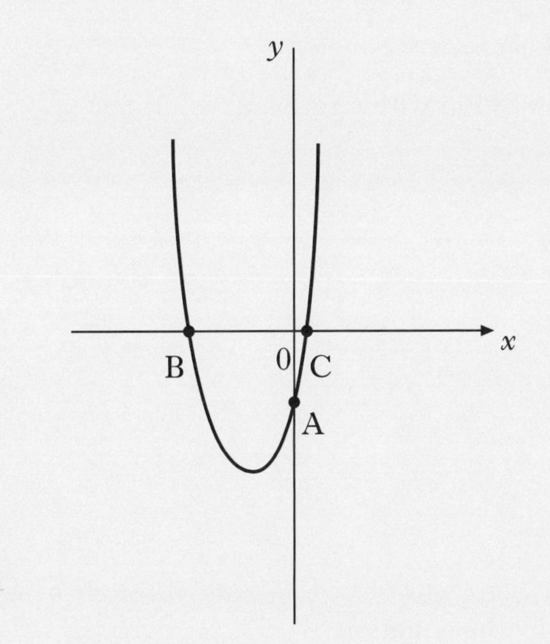

(*a*) Write down the coordinates of A. **1**

(*b*) Find the coordinates of B and C. **3**

(*c*) Calculate the minimum value of $4x^2 + 4x - 3$. **2**

9. A number pattern is shown below.

$$1^3 + 1 = (1 + 1)\,(1^2 - 1 + 1)$$
$$2^3 + 1 = (2 + 1)\,(2^2 - 2 + 1)$$
$$3^3 + 1 = (3 + 1)\,(3^2 - 3 + 1)$$

(*a*) Write down a similar expression for $7^3 + 1$. **1**

(*b*) Hence write down an expression for $n^3 + 1$. **1**

(*c*) Hence find an expression for $8p^3 + 1$. **2**

[Turn over

	KU	RE

10. Simplify

$$\frac{\sqrt{3}}{\sqrt{24}}.$$

Express your answer as a fraction with a rational denominator. **3** (KU)

11. The intensity of light, I, emerging after passing through a liquid with concentration, c, is given by the equation

$$I = \frac{20}{2^c} \qquad c \geq 0.$$

(*a*) Find the intensity of light when the concentration is 3. **1** (KU)

(*b*) Find the concentration of the liquid when the intensity is 10. **2** (KU)

(*c*) What is the maximum possible intensity? **3** (RE)

[*END OF QUESTION PAPER*]

C

2500/406

NATIONAL QUALIFICATIONS 2001

WEDNESDAY, 16 MAY
2.45 PM – 4.05 PM

MATHEMATICS
STANDARD GRADE
Credit Level
Paper 2

1 **You may use a calculator**.

2 Answer as many questions as you can.

3 Full credit will be given only where the solution contains appropriate working.

4 Square-ruled paper is provided.

FORMULAE LIST

The roots of $ax^2 + bx + c = 0$ are $x = \dfrac{-b \pm \sqrt{(b^2 - 4ac)}}{2a}$

Sine rule: $\dfrac{a}{\sin A} = \dfrac{b}{\sin B} = \dfrac{c}{\sin C}$

Cosine rule: $a^2 = b^2 + c^2 - 2bc \cos A$ or $\cos A = \dfrac{b^2 + c^2 - a^2}{2bc}$

Area of a triangle: Area $= \frac{1}{2}ab \sin C$

Volume of a cylinder: Volume $= \pi r^2 h$

Standard deviation: $s = \sqrt{\dfrac{\Sigma(x - \bar{x})^2}{n-1}} = \sqrt{\dfrac{\Sigma x^2 - (\Sigma x)^2/n}{n-1}}$, where n is the sample size.

	KU	RE

1.

How many chocpops will be eaten in the year 2001?

Give your answer in **scientific notation**. **2** (KU)

2. The price, in pence per litre, of petrol at 10 city garages is shown below.

84·2	84·4	85·1	83·9	81·0
84·2	85·6	85·2	84·9	84·8

(*a*) Calculate the mean and standard deviation of these prices. **3** (KU)

(*b*) In 10 rural garages, the petrol prices had a mean of 88·8 and a standard deviation of 2·4.

How do the rural prices compare with the city prices? **2** (RE)

3. In 1999, a house was valued at £90 000 and the contents were valued at £60 000.

The value of the house **appreciates** by 5% each year.

The value of the contents **depreciates** by 8% each year.

What will be the **total** value of the house **and** the contents in 2002? **3** (KU)

[Turn over

KU	RE

4. A water pipe runs between two buildings.

These are represented by the points A and B in the diagram below.

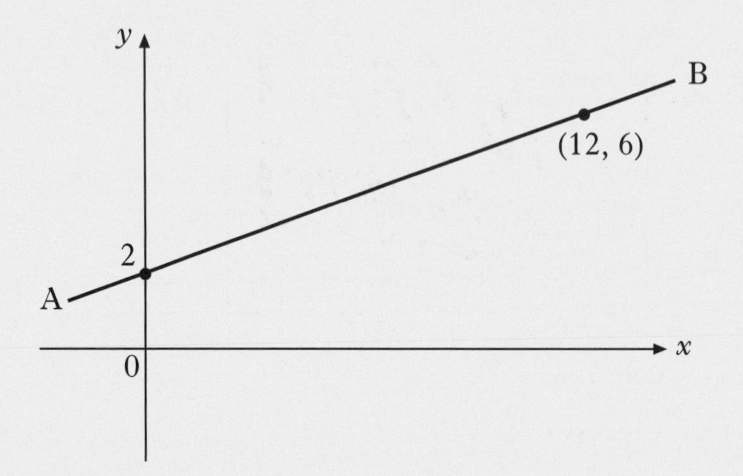

(*a*) Using the information in the diagram, show that the equation of the line AB is $3y - x = 6$. **3** (KU)

(*b*) An emergency outlet pipe has to be built across the main pipe. The line representing this outlet pipe has equation $4y + 5x = 46$.

Calculate the coordinates of the point on the diagram at which the outlet pipe will cut across the main water pipe. **4** (RE)

5. A cylindrical soft drinks can is 15 centimetres in height and 6·5 centimetres in diameter.

A new cylindrical can holds the same volume but has a reduced height of 12 centimetres.

What is the diameter of the new can?

Give your answer **to 1 decimal place**. **4** (RE)

KU	RE

6. Three radio masts, Kangaroo (K), Wallaby (W) and Possum (P) are situated in the Australian outback.

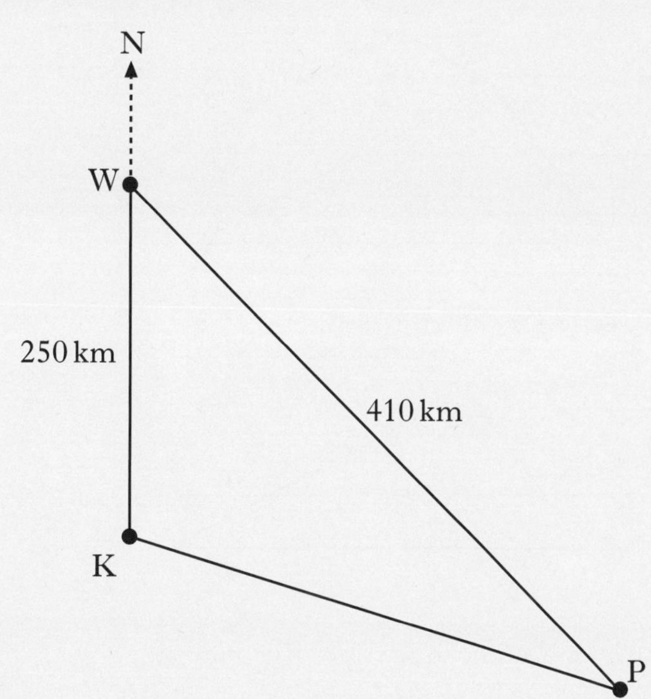

Kangaroo is 250 kilometres due south of Wallaby.

Wallaby is 410 kilometres from Possum.

Possum is on a bearing of 130° from Kangaroo.

Calculate the bearing of Possum from Wallaby.

Do not use a scale drawing. **4** (RE)

7. Solve **algebraically** the equation

$$\tan 40° = 2\sin x° + 1 \qquad 0 \leq x < 360.$$

3 (KU)

[Turn over

KU	RE

8. A metal door-stop is prism shaped, as shown.

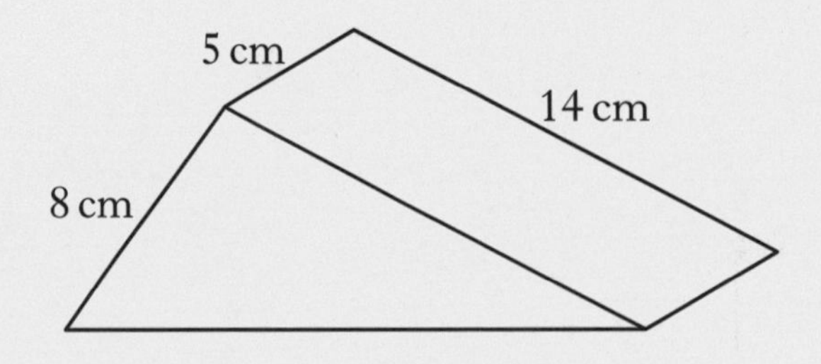

The uniform cross-section is shown below.

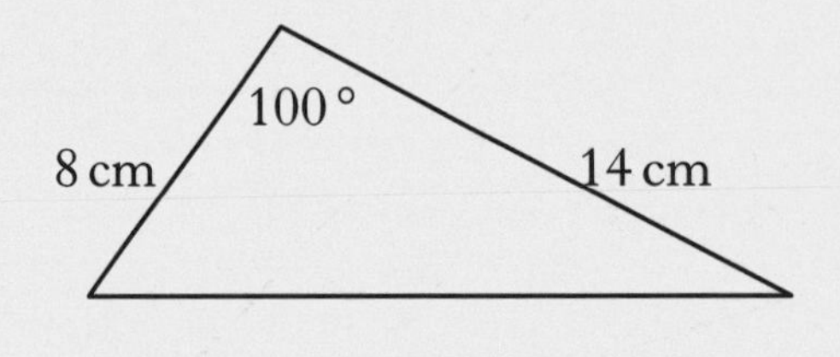

Find the volume of metal required to make the door-stop. **4** (KU)

9. The electrical resistance, R, of copper wire varies directly as its length, L metres, and inversely as the square of its diameter, d millimetres.

Two lengths of copper wire, A and B, have the same resistance.

Wire A has a diameter of 2 millimetres and a length of 3 metres.

Wire B has a diameter of 3 millimetres.

What is the length of wire B? **4** (RE)

KU | RE

10. Each leg of a folding table is prevented from opening too far by a metal bar.

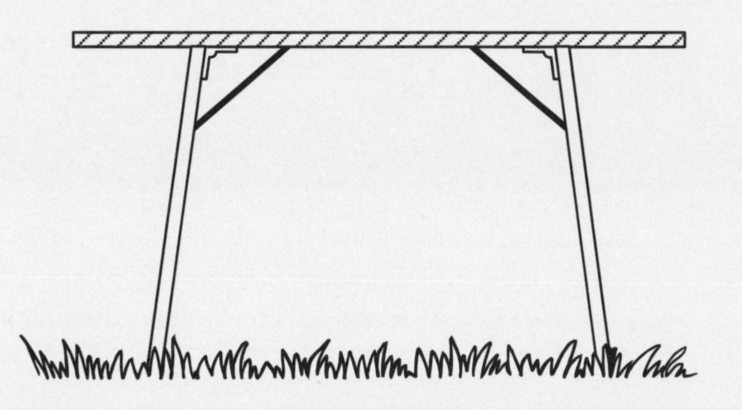

The metal bar is 21 centimetres long.

It is fixed to the table **top** 14 centimetres from the hinge and to the table **leg** 12 centimetres from the hinge.

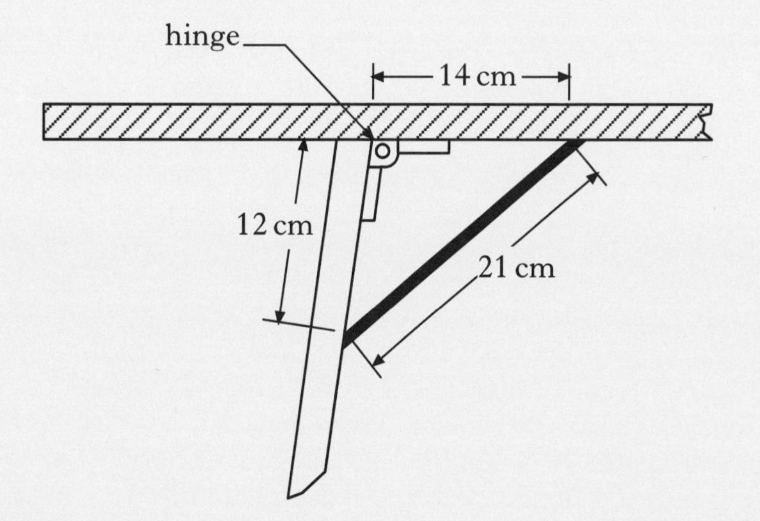

(*a*) Calculate the size of the obtuse angle which the table top makes with the leg. **3** (KU)

(*b*) Given that the table leg is 70 centimetres long, calculate the height of the table. **3** (RE)

[Turn over for Question 11 on *Page eight*

	KU	RE

11. A rectangular wall vent is 30 centimetres long and 20 centimetres wide.

It is to be enlarged by increasing **both** the length and the width by x centimetres.

(*a*) Write down the length of the new vent. — RE 1

(*b*) Show that the area, A square centimetres, of the new vent is given by

$$A = x^2 + 50x + 600.$$

RE 2

(*c*) The area of the new vent **must** be **at least** 40% more than the original area.

Find the **minimum** dimensions, to the nearest centimetre, of the new vent. — RE 5

[*END OF QUESTION PAPER*]

[BLANK PAGE]

C

2500/405

NATIONAL QUALIFICATIONS 2002

THURSDAY, 9 MAY
1.30 PM – 2.25 PM

MATHEMATICS
STANDARD GRADE
Credit Level
Paper 1
(Non-calculator)

1 **You may NOT use a calculator**.

2 Answer as many questions as you can.

3 Full credit will be given only where the solution contains appropriate working.

4 Square-ruled paper is provided.

LIB 2500/405 6/33570

FORMULAE LIST

The roots of $ax^2 + bx + c = 0$ are $x = \dfrac{-b \pm \sqrt{(b^2 - 4ac)}}{2a}$

Sine rule: $\dfrac{a}{\sin A} = \dfrac{b}{\sin B} = \dfrac{c}{\sin C}$

Cosine rule: $a^2 = b^2 + c^2 - 2bc \cos A$ or $\cos A = \dfrac{b^2 + c^2 - a^2}{2bc}$

Area of a triangle: Area $= \frac{1}{2}ab \sin C$

Standard deviation: $s = \sqrt{\dfrac{\sum(x - \bar{x})^2}{n-1}} = \sqrt{\dfrac{\sum x^2 - (\sum x)^2/n}{n-1}}$, where n is the sample size.

	KU	RE
1. Evaluate $7{\cdot}18 - 2{\cdot}1 \times 3$.	2	
2. Evaluate $1\frac{1}{8} \div \frac{3}{4}$.	2	
3. Solve the inequality $5 - x > 2(x + 1)$.	3	
4. Given that $f(x) = x^2 + 5x$, evaluate $f(-3)$.	2	
5. (*a*) Factorise $p^2 - 4q^2$.	1	
(*b*) Hence simplify $\frac{p^2 - 4q^2}{3p + 6q}$.	2	
6. $L = \frac{1}{2}(h - t)$. Change the subject of the formula to h.	2	

[Turn over

KU	RE

7. In triangle ABC,

AB = 4 units
AC = 5 units
BC = 6 units.

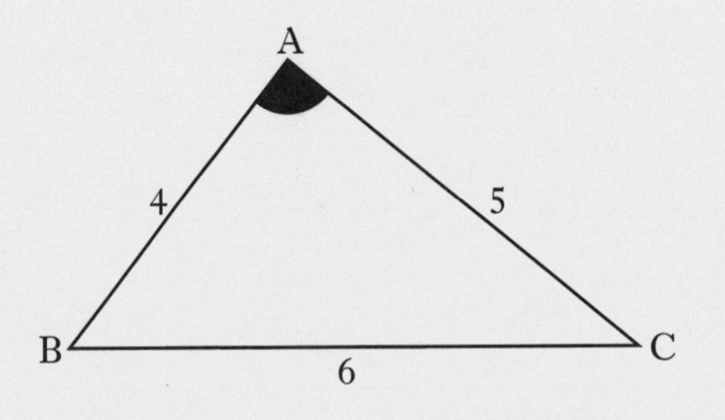

Show that $\cos A = \frac{1}{8}$. **3** (KU)

8. Fifteen medical centres each **handed out** a questionnaire to fifty patients.

The numbers who replied to each centre are shown below.

11	19	22	25	25
29	31	34	36	38
40	46	49	50	50

Also, they each **posted** the questionnaire to another fifty patients.

The numbers who replied to each centre are shown below.

15	15	21	22	23
25	26	31	33	34
37	39	41	46	46

Draw an appropriate statistical diagram to compare these two sets of data. **3** (RE)

9. Two functions are given below.

$$f(x) = x^2 + 2x - 1$$
$$g(x) = 5x + 3$$

Find the values of x for which $f(x) = g(x)$. **3** (RE)

	KU	RE

10. Simplify

$$\sqrt{27} + 2\sqrt{3}.$$

KU: 2

11. Express in its simplest form

$$y^8 \times (y^3)^{-2}.$$

KU: 2

12. The graph below shows the relationship between the history and geography marks of a class of students.

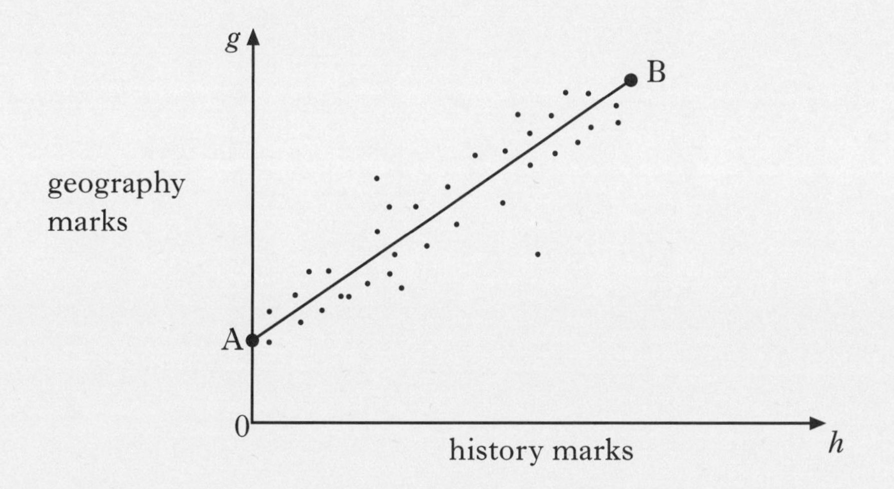

A best-fitting straight line, AB has been drawn.

Point A represents 0 marks for history and 12 marks for geography.
Point B represents 90 marks for history and 82 marks for geography.

Find the equation of the straight line AB in terms of h and g.

RE: 4

[Turn over for Question 13 on *Page six*

		KU	RE
13.	(*a*) 4 peaches and 3 grapefruit cost £1·30. Write down an algebraic equation to illustrate this.	1	
	(*b*) 2 peaches and 4 grapefruit cost £1·20. Write down an algebraic equation to illustrate this.	1	
	(*c*) Find the cost of 3 peaches and 2 grapefruit.		4

[*END OF QUESTION PAPER*]

C

2500/406

NATIONAL QUALIFICATIONS 2002

THURSDAY, 9 MAY
2.45 PM – 4.05 PM

MATHEMATICS
STANDARD GRADE
Credit Level
Paper 2

1 **You may use a calculator**.

2 Answer as many questions as you can.

3 Full credit will be given only where the solution contains appropriate working.

4 Square-ruled paper is provided.

LIB 2500/406 6/33570

FORMULAE LIST

The roots of $ax^2 + bx + c = 0$ are $x = \dfrac{-b \pm \sqrt{(b^2 - 4ac)}}{2a}$

Sine rule: $\dfrac{a}{\sin A} = \dfrac{b}{\sin B} = \dfrac{c}{\sin C}$

Cosine rule: $a^2 = b^2 + c^2 - 2bc \cos A$ or $\cos A = \dfrac{b^2 + c^2 - a^2}{2bc}$

Area of a triangle: Area $= \frac{1}{2}ab \sin C$

Standard deviation: $s = \sqrt{\dfrac{\sum(x - \overline{x})^2}{n-1}} = \sqrt{\dfrac{\sum x^2 - (\sum x)^2 / n}{n-1}}$, where n is the sample size.

	KU	RE
1. A spider weighs approximately $19{\cdot}06 \times 10^{-5}$ kilograms. A humming bird is 18 times heavier. Calculate the weight of the humming bird. Give your answer **in scientific notation**.	2	
2. A microwave oven is sold for £150. This price includes VAT at 17·5%. Calculate the price of the microwave oven **without** VAT.	3	
3. Solve the equation $2x^2 + 3x - 7 = 0.$ Give your answers **correct to 1 decimal place**.	4	

[Turn over

KU	RE

4. A TV signal is sent from a transmitter T, via a satellite S, to a village V, as shown in the diagram. The village is 500 kilometres from the transmitter.

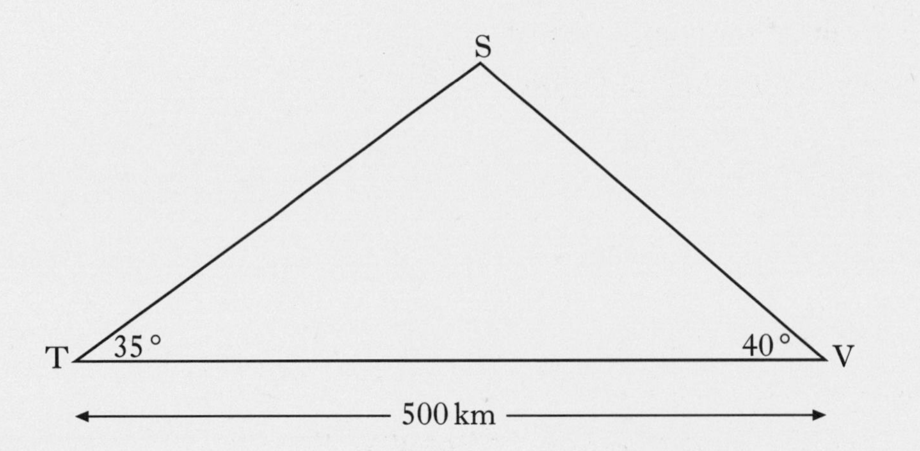

The signal is sent out at an angle of 35° and is received in the village at an angle of 40°.

Calculate the height of the satellite above the ground. **5** (RE)

5. A feeding trough, 4 metres long, is prism-shaped.

The uniform cross-section is made up of a rectangle and semi-circle as shown below.

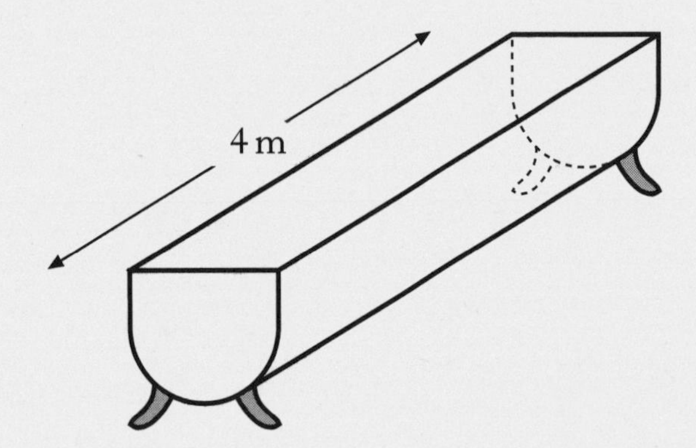

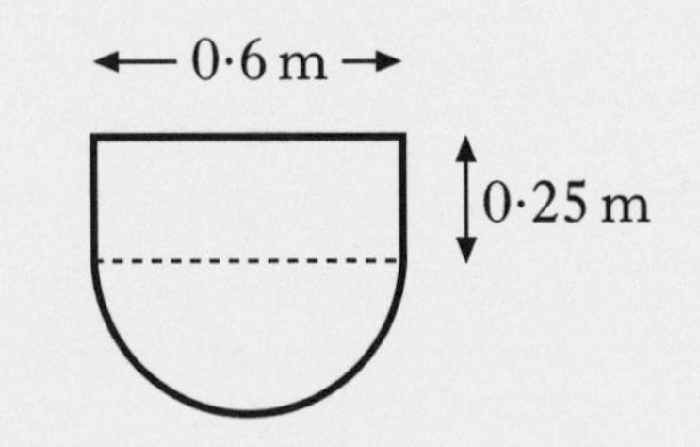

Find the volume of the trough, **correct to 2 significant figures**. **5** (KU)

	KU	RE

6. An oil tank has a circular cross-section of radius 2·1 metres.

It is filled to a depth of 3·4 metres.

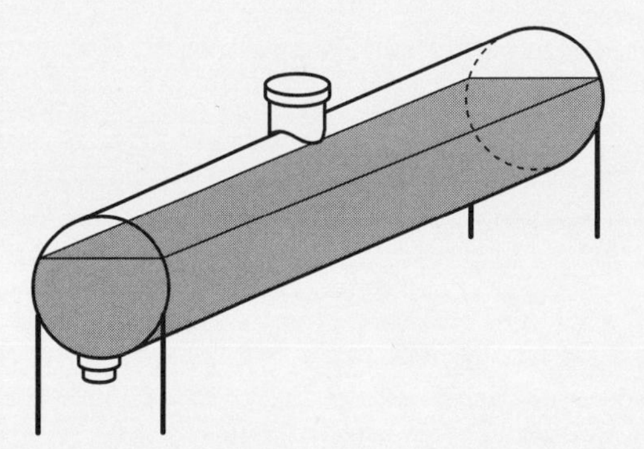

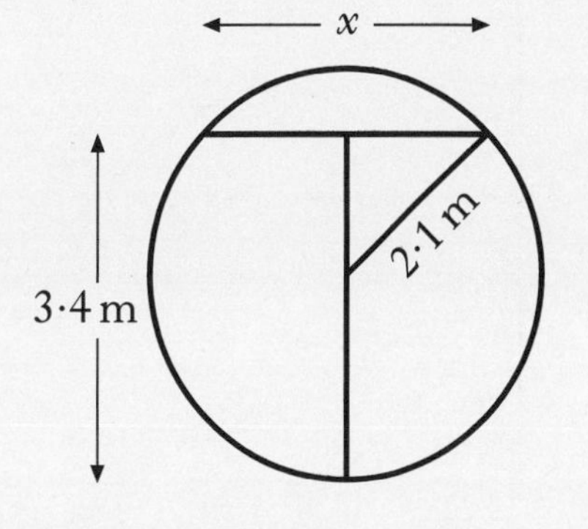

(*a*) Calculate x, the width in metres of the oil surface. **3** (KU)

(*b*) What other depth of oil would give the same surface width? **1** (RE)

7. A coffee shop blends its own coffee and sells it in one-kilogram tins.

One blend consists of two kinds of coffee, Brazilian and Colombian, in the ratio 2 : 3.

The shop has 20 kilograms of Brazilian and 25 kilograms of Colombian in stock.

What is the **maximum** number of one-kilogram tins of this blend which can be made? **3** (RE)

[Turn over

KU | RE

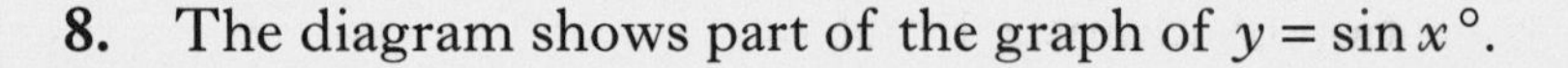

8. The diagram shows part of the graph of $y = \sin x°$.

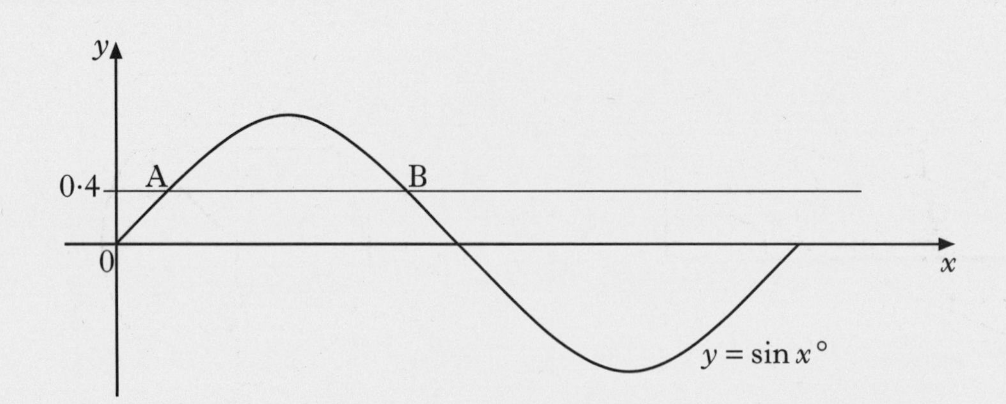

The line $y = 0{\cdot}4$ is drawn and cuts the graph of $y = \sin x°$ at A and B.

Find the x-coordinates of A and B. **3**

9. Esther has a new mobile phone and considers the following daily rates.

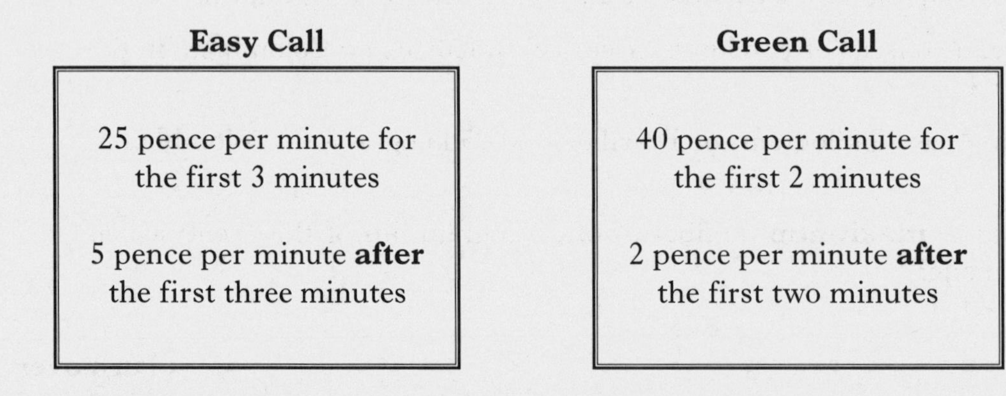

(*a*) For Easy Call, find the cost of ten minutes in a day. **1**

(*b*) For Easy Call, find a formula for the cost of "m" minutes in a day, $m > 3$. **1**

(*c*) For Green Call, find a formula for the cost of "m" minutes in a day, $m > 2$. **1**

(*d*) Green Call claims that its system is cheaper.

Find **algebraically** the least number of minutes (to the nearest minute) which must be used each day for this claim to be true. **3**

	KU	RE

10. A weight on the end of a string is spun in a circle on a smooth table.

The tension, T, in the string varies directly as the square of the speed, v, and inversely as the radius, r, of the circle.

(*a*) Write down a formula for T in terms of v and r. **1** (KU)

(*b*) The speed of the weight is multiplied by 3 and the radius of the string is halved.

What happens to the tension in the string? **2** (RE)

11. (*a*) Solve the equation

$$2^n = 32.$$

1 (KU)

(*b*) A sequence of numbers can be grouped and added together as shown.

The sum of 2 numbers:	$(1 + 2)$	$=$	$4 - 1$
The sum of 3 numbers:	$(1 + 2 + 4)$	$=$	$8 - 1$
The sum of 4 numbers:	$(1 + 2 + 4 + 8)$	$=$	$16 - 1$

Find a **similar** expression for the sum of 5 numbers. **1** (RE)

(*c*) Find a formula for the sum of the first n numbers of this sequence. **2** (RE)

[Turn over for Question 12 on *Page eight*

KU	RE

12. A metal beam, AB, is 6 metres long.

It is hinged at the top, P, of a vertical post 1 metre high.

When B touches the ground, A is 1·5 metres above the ground, as shown in Figure 1.

Figure 1

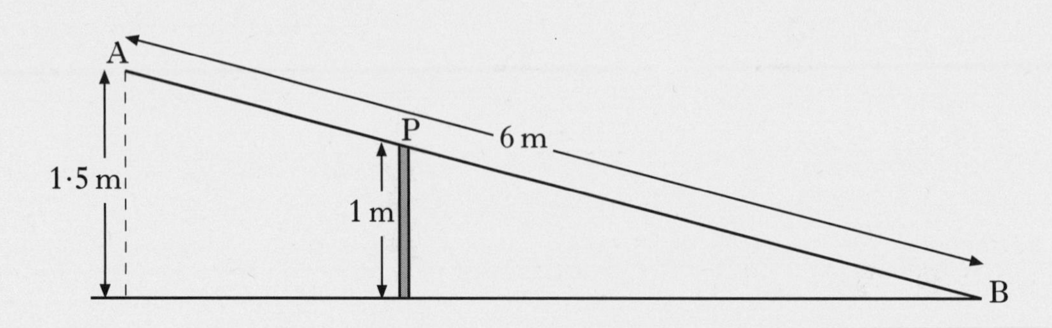

When A comes down to the ground, B rises, as shown in Figure 2.

Figure 2

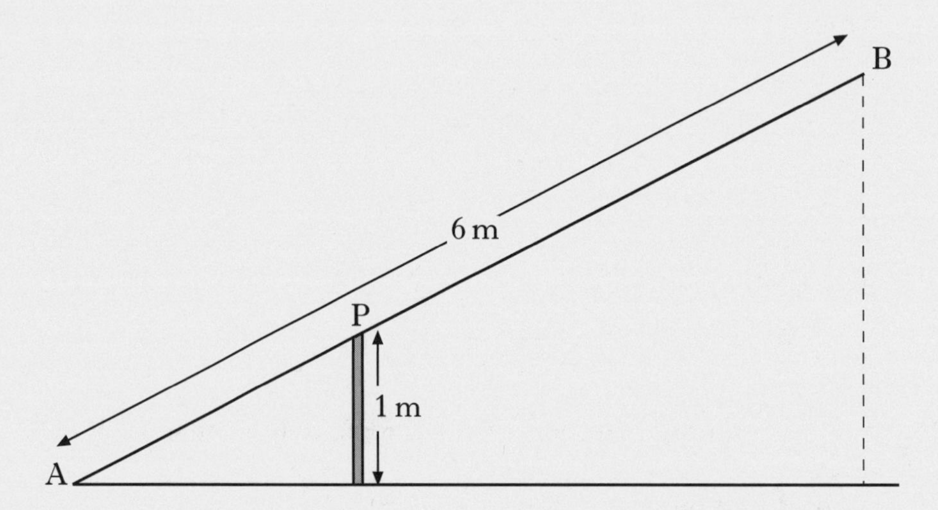

By calculating the length of AP, or otherwise, find the height of B above the ground.

Do not use a scale drawing. **5**

[*END OF QUESTION PAPER*]

[BLANK PAGE]

C

2500/405

NATIONAL
QUALIFICATIONS
2003

THURSDAY, 8 MAY
1.30 PM – 2.25 PM

MATHEMATICS
STANDARD GRADE
Credit Level
Paper 1
(Non-calculator)

1 **You may NOT use a calculator**.

2 Answer as many questions as you can.

3 Full credit will be given only where the solution contains appropriate working.

4 Square-ruled paper is provided.

FORMULAE LIST

The roots of $ax^2 + bx + c = 0$ are $x = \dfrac{-b \pm \sqrt{(b^2 - 4ac)}}{2a}$

Sine rule: $\dfrac{a}{\sin A} = \dfrac{b}{\sin B} = \dfrac{c}{\sin C}$

Cosine rule: $a^2 = b^2 + c^2 - 2bc \cos A$ or $\cos A = \dfrac{b^2 + c^2 - a^2}{2bc}$

Area of a triangle: Area $= \frac{1}{2}ab \sin C$

Standard deviation: $s = \sqrt{\dfrac{\sum(x - \bar{x})^2}{n-1}} = \sqrt{\dfrac{\sum x^2 - (\sum x)^2 / n}{n-1}}$, where n is the sample size.

		KU	RE
1.	Evaluate $5{\cdot}04 + 8{\cdot}4 \div 7$.	2	
2.	Evaluate $\frac{2}{7}(1\frac{3}{4} + \frac{3}{8})$.	2	
3.	Simplify $3(2x - 4) - 4(3x + 1)$.	3	
4.	$f(x) = 7 - 4x$		
	(*a*) Evaluate $f(-2)$.	1	
	(*b*) Given that $f(t) = 9$, find t.	2	
5.	Factorise $2x^2 - 7x - 15$.	2	

[Turn over

	KU	RE

6. In the diagram below, A is the point (–1, –7) and B is the point (4, 3).

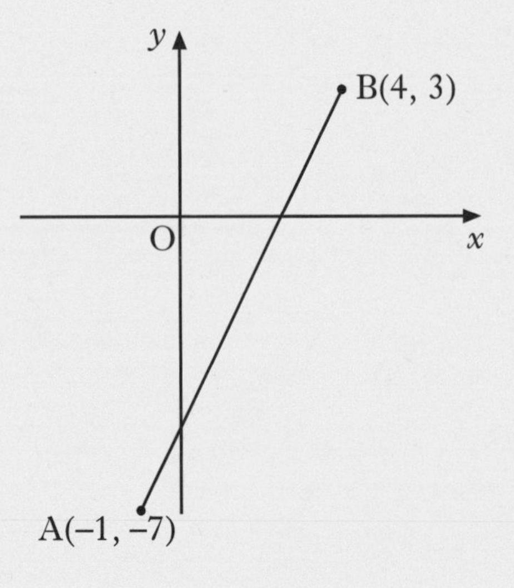

(*a*) Find the gradient of the line AB. — KU 1

(*b*) AB cuts the y-axis at the point (0, –5).
Write down the equation of the line AB. — KU 1

(*c*) The point $(3k, k)$ lies on AB.
Find the value of k. — RE 2

7. Andrew and Doreen each book in at the Sleepwell Lodge.

(*a*) Andrew stays for 3 nights and has breakfast on 2 mornings.
His bill is £145.
Write down an algebraic equation to illustrate this. — KU 1

(*b*) Doreen stays for 5 nights and has breakfast on 3 mornings.
Her bill is £240.
Write down an algebraic equation to illustrate this. — KU 1

(*c*) Find the cost of one breakfast. — RE 3

	KU	RE
8. A mini lottery game uses **red**, **green**, **blue** and **yellow** balls. There are 10 of **each** colour, numbered from 1 to 10. The balls are placed in a drum and one is drawn out.		
(*a*) What is the probability that it is a **6**?	1	
(*b*) What is the probability that it is a **yellow 6**?	1	
9. A random check is carried out on the contents of a number of matchboxes. A summary of the results is shown in the boxplot below.		

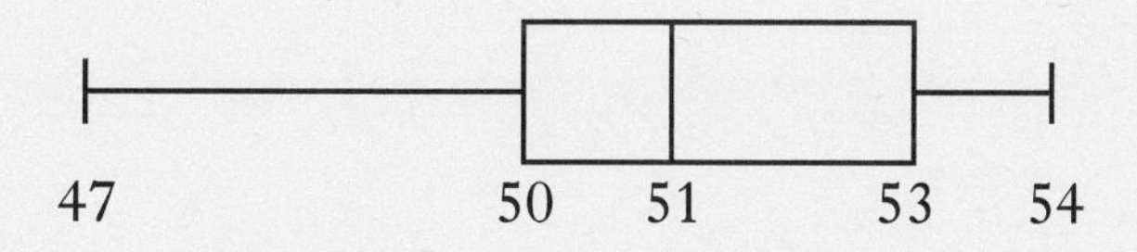

	KU	RE
What percentage of matchboxes contains fewer than 50 matches?		1
10. School theatre visits are arranged for parents, teachers and pupils. The ratio of parents to teachers to pupils **must** be 1 : 3 : 15.		
(*a*) 45 pupils want to go to the theatre. How many teachers must accompany them?	1	
(*b*) The theatre gives the school 100 tickets for a play. What is the maximum number of pupils who can go to the play?		3

[Turn over

	KU	RE
11. Using the sequence		
$$1, \; 3, \; 5, \; 7, \; 9, \ldots.$$		
(*a*) Find S_3, the sum of the first 3 numbers.		1
(*b*) Find S_n, the sum of the first n numbers.		2
(*c*) Hence or otherwise, find the $(n+1)^{th}$ term of the sequence.		2
12. (*a*) Evaluate		
$$8^{\frac{2}{3}}.$$	2	
(*b*) Simplify		
$$\frac{\sqrt{24}}{\sqrt{2}}.$$	2	

KU	RE

13. A rectangular clipboard has a triangular plastic pocket attached as shown in Figure 1.

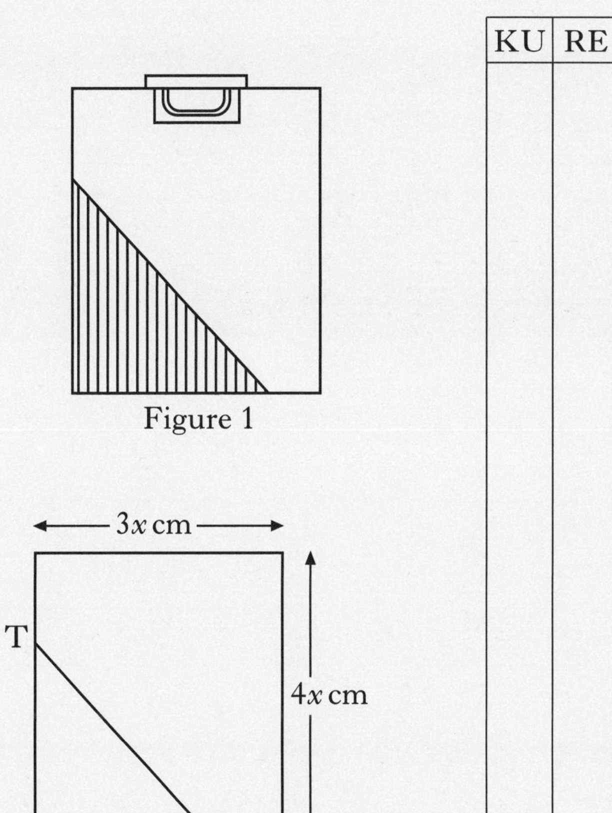

Figure 1

The pocket is attached along edges TD and DB as shown in Figure 2.

B is x centimetres from the corner C.

Figure 2

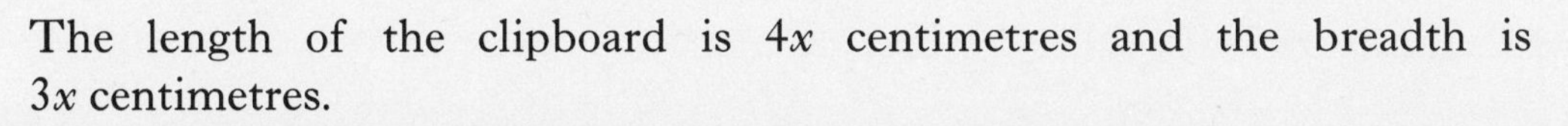

The length of the clipboard is $4x$ centimetres and the breadth is $3x$ centimetres.

The area of the pocket is a quarter of the area of the clipboard.

Find, in terms of x, the length of TD. **4**

[*END OF QUESTION PAPER*]

[BLANK PAGE]

C

2500/406

NATIONAL
QUALIFICATIONS
2003

THURSDAY, 8 MAY
2.45 PM – 4.05 PM

MATHEMATICS
STANDARD GRADE
Credit Level
Paper 2

1 **You may use a calculator**.

2 Answer as many questions as you can.

3 Full credit will be given only where the solution contains appropriate working.

4 Square-ruled paper is provided.

FORMULAE LIST

The roots of $ax^2 + bx + c = 0$ are $x = \dfrac{-b \pm \sqrt{(b^2 - 4ac)}}{2a}$

Sine rule: $\dfrac{a}{\sin A} = \dfrac{b}{\sin B} = \dfrac{c}{\sin C}$

Cosine rule: $a^2 = b^2 + c^2 - 2bc \cos A$ or $\cos A = \dfrac{b^2 + c^2 - a^2}{2bc}$

Area of a triangle: Area $= \frac{1}{2}ab \sin C$

Standard deviation: $s = \sqrt{\dfrac{\sum(x - \bar{x})^2}{n-1}} = \sqrt{\dfrac{\sum x^2 - (\sum x)^2/n}{n-1}}$, where n is the sample size.

	KU	RE

1. Bacteria in a test-tube increase at the rate of 0·6% per hour.

At 12 noon, there are 5000 bacteria.

At 3pm, how many bacteria will be present?

Give your answer **to 3 significant figures**. — KU 4

2. Fiona checks out the price of a litre of milk in several shops.

The prices in pence are:

49 44 41 52 47 43.

(*a*) Find the mean price of a litre of milk. — KU 1

(*b*) Find the standard deviation of the prices. — KU 2

(*c*) Fiona also checks out the price of a kilogram of sugar in the same shops and finds that the standard deviation of the prices is 2·6.

Make one valid comparison between the two sets of prices. — RE 1

3. Two yachts leave from harbour H.

Yacht A sails on a bearing of 072° for 30 kilometres and stops.

Yacht B sails on a bearing of 140° for 50 kilometres and stops.

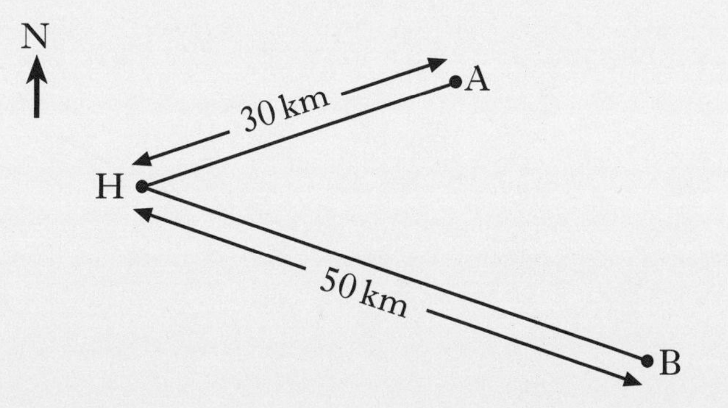

How far apart are the two yachts when they have both stopped?

Do not use a scale drawing. — RE 4

[Turn over

KU	RE

4. A mug is in the shape of a cylinder with diameter 10 centimetres and height 14 centimetres.

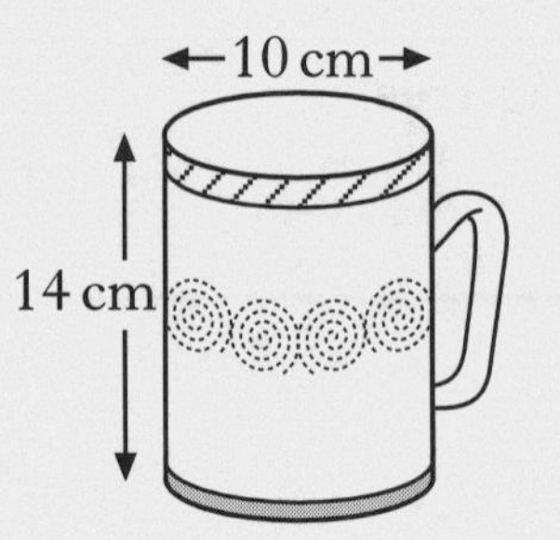

(*a*) Calculate the volume of the mug. **2** (KU)

(*b*) 600 millilitres of coffee are poured in.

Calculate the depth of the coffee in the cup. **3** (RE)

5. The number of diagonals, d, in a polygon with n sides is given by the formula

$$d = \frac{n(n-3)}{2}.$$

A polygon has 20 diagonals.

How many sides does it have? **4** (RE)

6. In the diagram,

Angle STV = 34°

Angle VSW = 25°

Angle SVT = Angle SWV = 90°

ST = 13·1 centimetres.

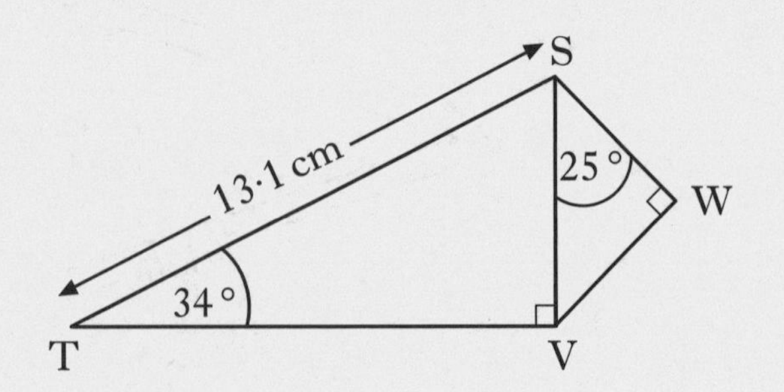

Calculate the length of SW. **4** (KU)

	KU	RE

7. The area of triangle ABC is 38 square centimetres.

AB is 9 centimetres and BC is 14 centimetres.

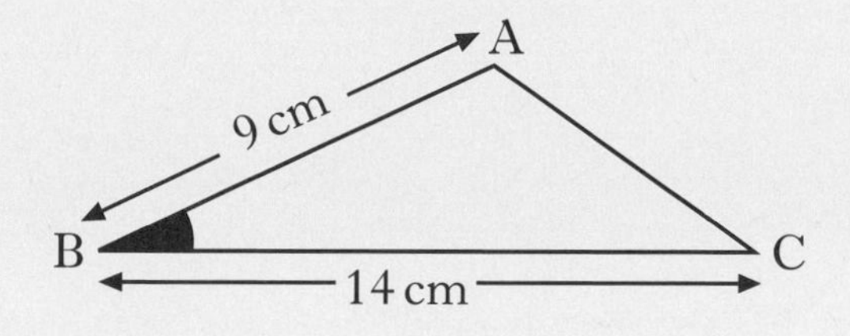

Calculate the size of the acute angle ABC. — RE: 3

8. The diagram below shows part of the graph of a quadratic function, with equation of the form $y = k(x - a)(x - b)$.

The graph cuts the y-axis at $(0, -6)$ and the x-axis at $(-1, 0)$ and $(3, 0)$.

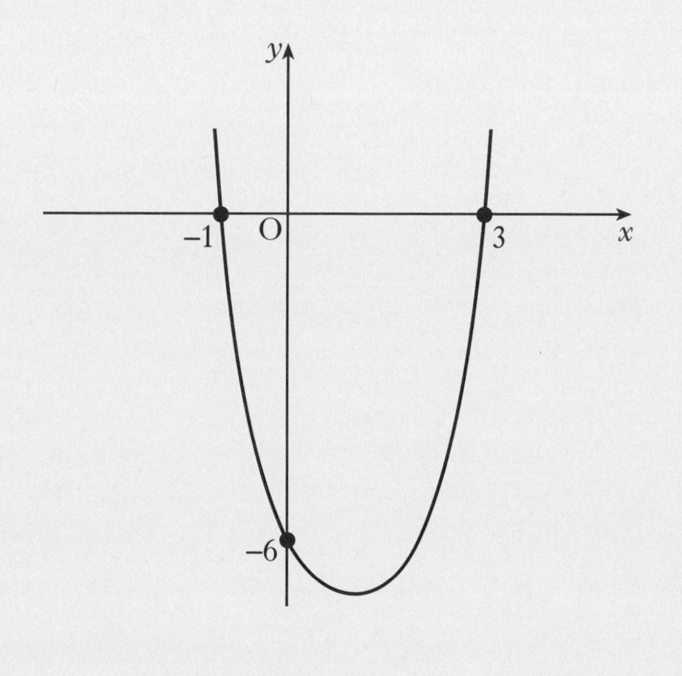

(*a*) Write down the values of a and b. — KU: 2

(*b*) Calculate the value of k. — KU: 2

(*c*) Find the coordinates of the minimum turning point of the function. — RE: 2

[Turn over

KU	RE

9. Two perfume bottles are mathematically similar in shape.

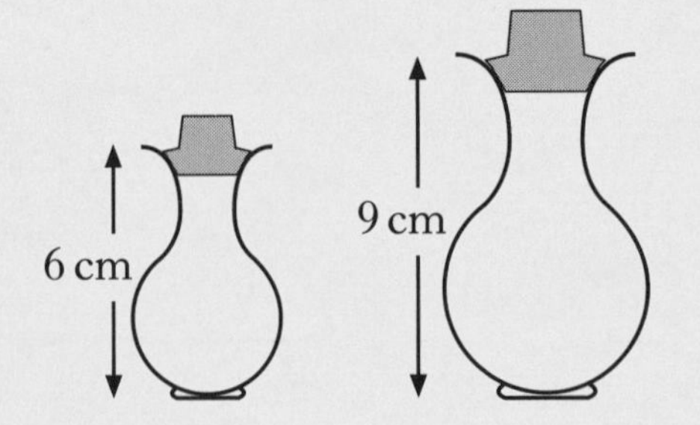

The smaller one is 6 centimetres high and holds 30 millilitres of perfume.

The larger one is 9 centimetres high.

What volume of perfume will the larger one hold? **3** (KU)

10. A sheep shelter is part of a cylinder as shown in Figure 1.

It is 6 metres wide and 2 metres high.

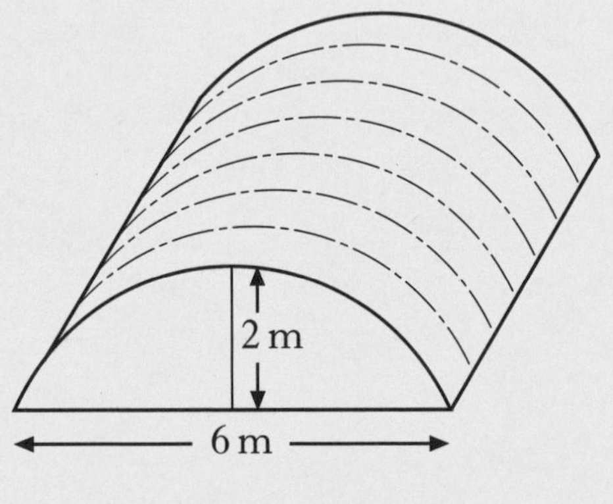

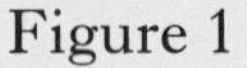

Figure 1

The cross-section of the shelter is a segment of a circle with centre O, as shown in Figure 2.

OB is the radius of the circle.

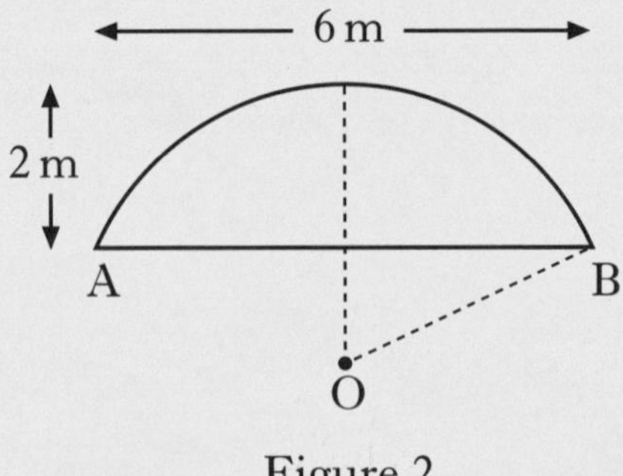

Figure 2

Calculate the length of OB. **4** (RE)

	KU	RE
11. (*a*) A driver travels from A to B, a distance of x kilometres, at a constant speed of 75 kilometres per hour. Find the time taken for this journey in terms of x.	**1**	
(*b*) The time for the journey from B to A is $\frac{x}{50}$ hours. Hence calculate the driver's average speed for the whole journey.		**4**

[*END OF QUESTION PAPER*]

[BLANK PAGE]

[BLANK PAGE]

C

2500/405

NATIONAL
QUALIFICATIONS
2004

FRIDAY, 7 MAY
1.30 PM – 2.25 PM

MATHEMATICS
STANDARD GRADE
Credit Level
Paper 1
(Non-calculator)

1 **You may NOT use a calculator**.

2 Answer as many questions as you can.

3 Full credit will be given only where the solution contains appropriate working.

4 Square-ruled paper is provided.

LIB 2500/405 6/33320

FORMULAE LIST

The roots of $ax^2 + bx + c = 0$ are $x = \dfrac{-b \pm \sqrt{(b^2 - 4ac)}}{2a}$

Sine rule: $\dfrac{a}{\sin A} = \dfrac{b}{\sin B} = \dfrac{c}{\sin C}$

Cosine rule: $a^2 = b^2 + c^2 - 2bc \cos A$ or $\cos A = \dfrac{b^2 + c^2 - a^2}{2bc}$

Area of a triangle: Area $= \frac{1}{2}ab\sin C$

Standard deviation: $s = \sqrt{\dfrac{\sum(x - \bar{x})^2}{n-1}} = \sqrt{\dfrac{\sum x^2 - (\sum x)^2/n}{n-1}}$, where n is the sample size.

KU	RE

1. Evaluate

$$6{\cdot}2 - (4{\cdot}53 - 1{\cdot}1).$$

2 (KU)

2. Evaluate $\frac{2}{5}$ of $3\frac{1}{2} + \frac{4}{5}$

3 (KU)

3. $A = 2x^2 - y^2$.

Calculate the value of A when $x = 3$ and $y = -4$.

2 (KU)

4. Simplify $\frac{3}{m} + \frac{4}{(m+1)}$

3 (KU)

5. The average monthly temperature in a holiday resort was recorded in degrees Celsius (°C).

Month	Jan	Feb	Mar	Apr	May	June	July	Aug	Sept	Oct	Nov	Dec
Average Temperature (°C)	1	8	8	10	15	22	23	24	20	14	9	4

Draw a suitable statistical diagram to illustrate the median and the quartiles of this data.

4 (RE)

[Turn over

KU	RE

6. Marmalade is on special offer.

Each jar on special offer contains 12·5% more than the standard jar.

A jar on special offer contains 450 g of marmalade.

How much does the standard jar contain? **3** (KU)

7. John's school sells 1200 tickets for a raffle.

John buys 15 tickets.

John's church sells 1800 tickets for a raffle.

John buys 20 tickets.

In which raffle has he a better chance of winning the first prize?

Show clearly all your working. **3** (RE)

KU | RE

8. 7, −2, 5, 3, 8

In the sequence above, each term after the first two terms is the sum of the previous two terms.

For example: 3rd term = 5 = 7 + (−2)

(*a*) A sequence follows the above rule.

The first term is x and the second term is y.

The fifth term is 5.

x, y, –, –, 5

Show that $2x + 3y = 5$ — RE **2**

(*b*) Using the same x and y, another sequence follows the above rule.

The first term is y and the second term is x.

The sixth term is 17.

y, x, –, –, –, 17.

Write down another equation in x and y. — RE **2**

(*c*) Find the values of x and y. — RE **3**

9. The graph of $y = a \cos bx°$, $0 \leq x \leq 90$, is shown below.

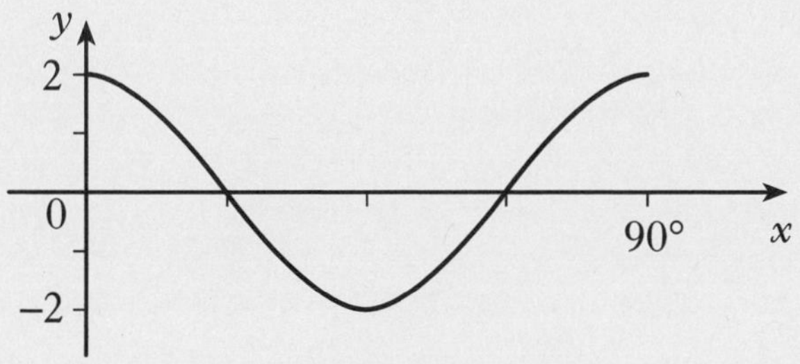

Write down the values of a and b. — KU **2**

[Turn over for Questions 10, 11 and 12 on *Page six*

	KU	RE
10. Two variables x and y are connected by the relationship $y = ax + b$. Sketch a possible graph of y against x to illustrate this relationship when a and b are each less than zero.		3
11. (*a*) Simplify $2\sqrt{75}$	2	
(*b*) Evaluate $2^0 + 3^{-1}$.	2	
12. A piece of gold wire 10 centimetres long is made into a circle.		

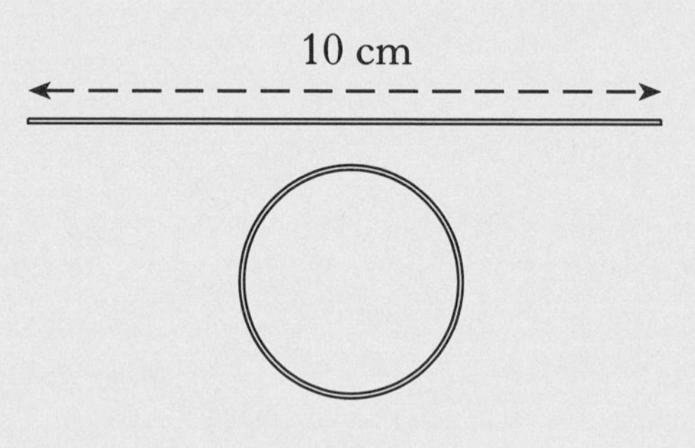

The circumference of the circle is equal to the length of the wire.

Show that the area of the circle is **exactly** $\frac{25}{\pi}$ square centimetres. — RE: 4

[*END OF QUESTION PAPER*]

C

2500/406

NATIONAL QUALIFICATIONS 2004

FRIDAY, 7 MAY
2.45 PM – 4.05 PM

MATHEMATICS STANDARD GRADE
Credit Level
Paper 2

1 **You may use a calculator**.

2 Answer as many questions as you can.

3 Full credit will be given only where the solution contains appropriate working.

4 Square-ruled paper is provided.

FORMULAE LIST

The roots of $ax^2 + bx + c = 0$ are $x = \dfrac{-b \pm \sqrt{(b^2 - 4ac)}}{2a}$

Sine rule: $\dfrac{a}{\sin A} = \dfrac{b}{\sin B} = \dfrac{c}{\sin C}$

Cosine rule: $a^2 = b^2 + c^2 - 2bc \cos A$ or $\cos A = \dfrac{b^2 + c^2 - a^2}{2bc}$

Area of a triangle: Area $= \frac{1}{2}ab\sin C$

Standard deviation: $s = \sqrt{\dfrac{\sum(x - \bar{x})^2}{n-1}} = \sqrt{\dfrac{\sum x^2 - (\sum x)^2/n}{n-1}}$, where n is the sample size.

	KU	RE

1. Radio signals travel at a speed of 3×10^8 metres per second.

A radio signal from Earth to a space probe takes 8 hours.

What is the distance from Earth to the probe?

Give your answer **in scientific notation**. **4** (KU)

2. A tank which holds 100 litres of water has a leak.

After 150 minutes, there is no water left in the tank.

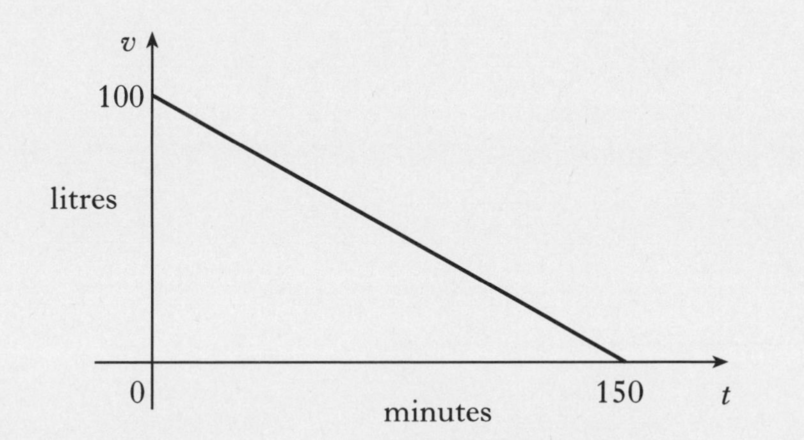

The above graph represents the volume of water (v litres) against time (t minutes).

(*a*) Find the equation of the line in terms of v and t. **3** (KU)

(*b*) How many minutes does it take for the container to lose 30 litres of water? **3** (RE)

3. Bottles of juice should contain 50 millilitres.

The contents of 7 bottles are checked in a random sample.

The actual volumes in millilitres are as shown below.

52, 50, 51, 49, 52, 53, 50

Calculate the mean and standard deviation of the sample. **4** (KU)

	KU	RE
4. 250 milligrams of a drug are given to a patient at 12 noon. The amount of the drug in the bloodstream decreases by 20% every hour. How many milligrams of the drug are in the bloodstream at 3pm?	**3**	
5. A helicopter, at point H, hovers between two aircraft carriers at points A and B which are 1500 metres apart. From carrier A, the angle of elevation of the helicopter is 50°. From carrier B, the angle of elevation of the helicopter is 55°. Calculate the distance from the helicopter to the nearer carrier.		**4**

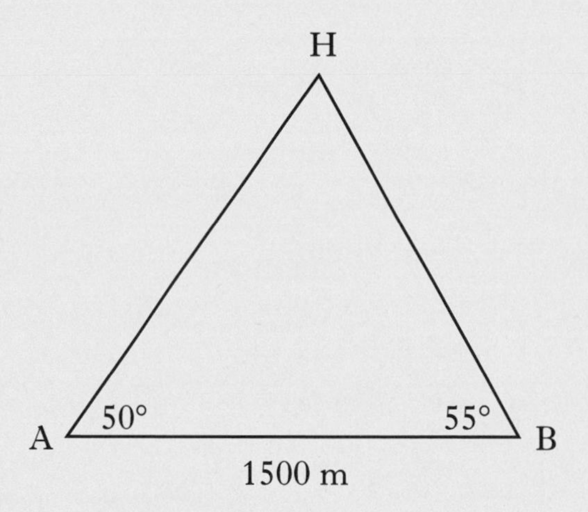

KU	RE

6. The diagram below shows a spotlight at point S, mounted 10 metres directly above a point P at the front edge of a stage.

The spotlight swings 45° from the vertical to illuminate another point Q, also at the front edge of the stage.

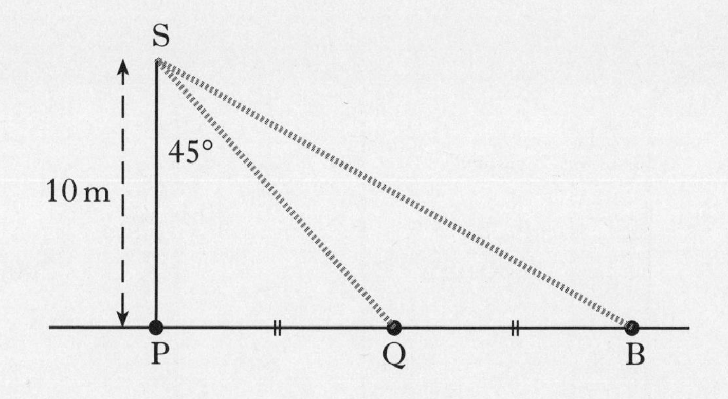

Through how many **more** degrees must the spotlight swing to illuminate a point B, where Q is the mid-point of PB? **5** (RE)

7. A square trapdoor of side 80 centimetres is held open by a rod as shown.

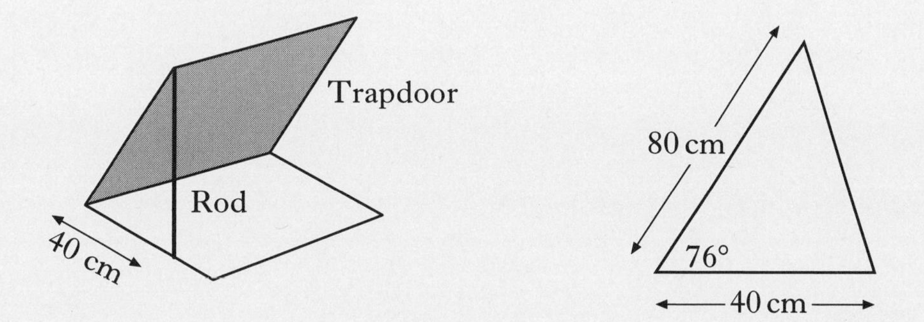

The rod is attached to a corner of the trapdoor and placed 40 centimetres along the edge of the opening.

The angle between the trapdoor and the opening is 76°.

Calculate the length of the rod to **2 significant figures**. **4** (KU)

[Turn over

KU	RE

8. The curved part of a doorway is an arc of a circle with radius 500 millimetres and centre C.

The height of the doorway to the top of the arc is 2000 millimetres.

The vertical edge of the doorway is 1800 millimetres.

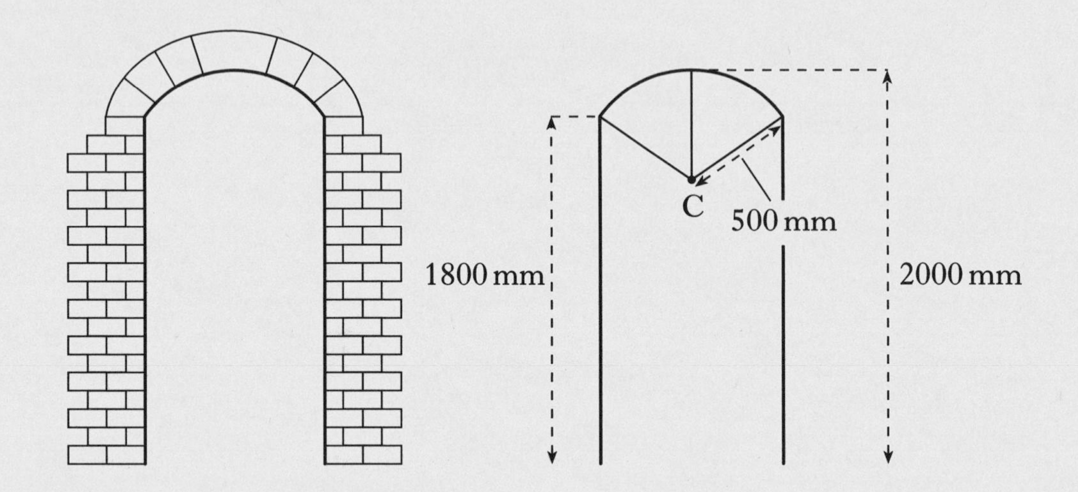

Calculate the width of the doorway. **5** (RE)

9. A gift box, 8 centimetres high, is prism shaped.

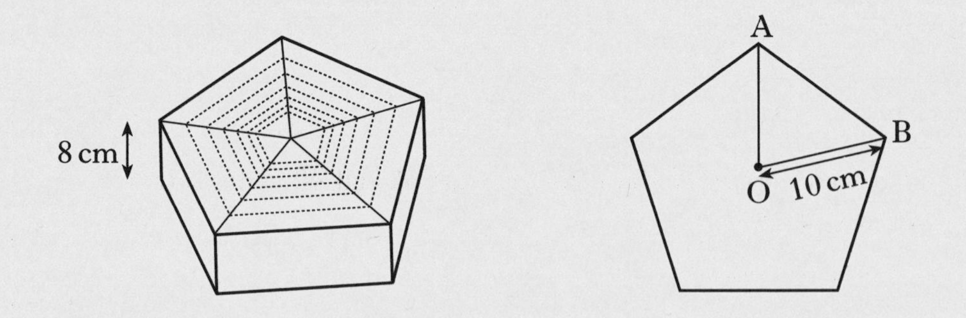

The uniform cross-section is a regular pentagon.

Each vertex of the pentagon is 10 centimetres from the centre O.

Calculate the volume of the box. **5** (KU)

KU | RE

10. Solve **algebraically** the equation

$$4 \sin x° + 1 = -2 \qquad 0 \leq x < 360.$$

KU: 3

11. A rectangular lawn has a path, 1 metre wide, on 3 sides as shown.

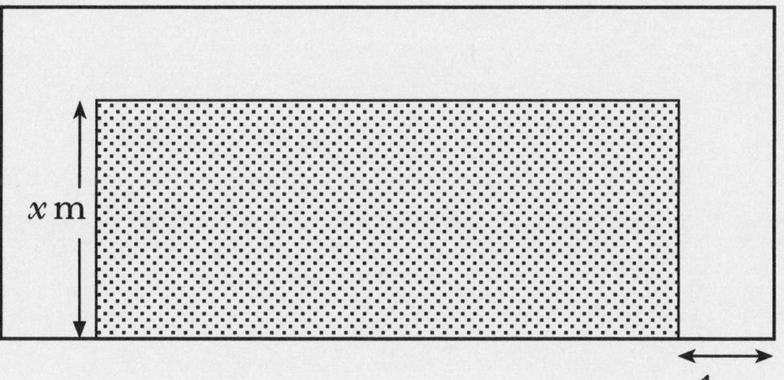

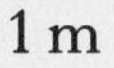

The breadth of the lawn is x metres.

The length of the lawn is three times its breadth.

The area of the lawn equals the area of the path.

(*a*) Show that $3x^2 - 5x - 2 = 0$. RE: 3

(*b*) Hence find the **length** of the lawn. RE: 4

[*END OF QUESTION PAPER*]

[BLANK PAGE]

2005 | Credit

[BLANK PAGE]

C

2500/405

NATIONAL
QUALIFICATIONS
2005

FRIDAY, 6 MAY
1.30 PM – 2.25 PM

MATHEMATICS
STANDARD GRADE
Credit Level
Paper 1
(Non-calculator)

1 **You may NOT use a calculator**.

2 Answer as many questions as you can.

3 Full credit will be given only where the solution contains appropriate working.

4 Square-ruled paper is provided.

FORMULAE LIST

The roots of $ax^2 + bx + c = 0$ are $x = \dfrac{-b \pm \sqrt{(b^2 - 4ac)}}{2a}$

Sine rule: $\dfrac{a}{\sin A} = \dfrac{b}{\sin B} = \dfrac{c}{\sin C}$

Cosine rule: $a^2 = b^2 + c^2 - 2bc \cos A$ or $\cos A = \dfrac{b^2 + c^2 - a^2}{2bc}$

Area of a triangle: Area $= \frac{1}{2}ab \sin C$

Standard deviation: $s = \sqrt{\dfrac{\sum(x - \overline{x})^2}{n-1}} = \sqrt{\dfrac{\sum x^2 - (\sum x)^2/n}{n-1}}$, where n is the sample size.

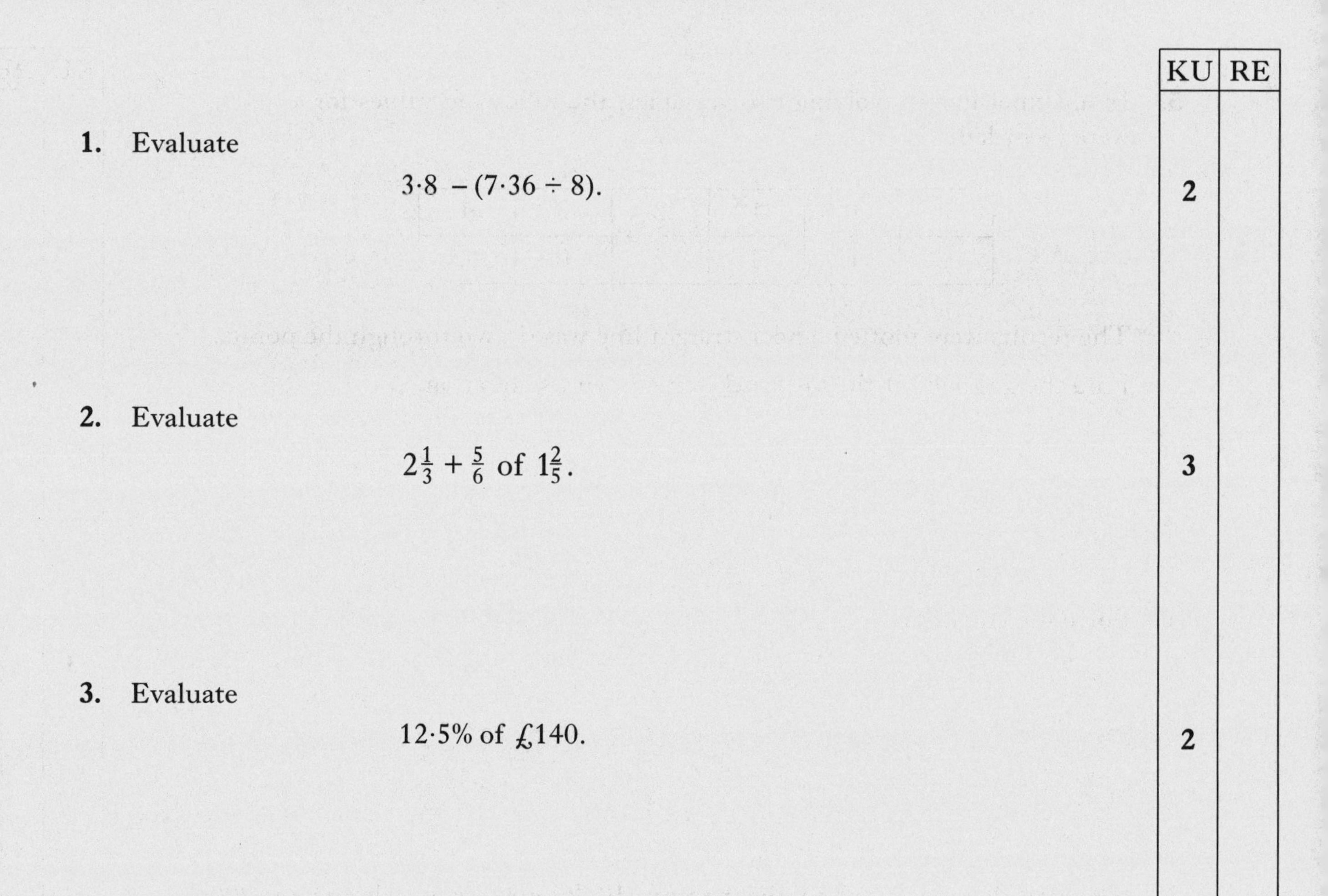

	KU	RE
1. Evaluate $3{\cdot}8 - (7{\cdot}36 \div 8)$.	2	
2. Evaluate $2\frac{1}{3} + \frac{5}{6}$ of $1\frac{2}{5}$.	3	
3. Evaluate 12·5% of £140.	2	

4. Two identical dice are rolled simultaneously.

Find the probability that the total score on adding both numbers will be greater than 7 but less than 10. — KU 2

[Turn over

KU | RE

5. In an experiment involving two variables, the following values for x and y were recorded.

x	0	1	2	3	4
y	6	4	2	0	–2

The results were plotted, and a straight line was drawn through the points.

Find the gradient of the line **and** write down its equation. 3

6. Solve the equation

$$\frac{2}{x}+1=6.$$

3

7. The speeds (measured to the nearest 10 kilometres per hour) of 200 vehicles are recorded as shown.

Speed (km/hr)	30	40	50	60	70	80	90	100	110
Frequency	1	4	9	14	38	47	51	32	4

Construct a cumulative frequency table and hence find the median for this data. 3

8. A number pattern is given below.

1st term: $2^2 - 0^2$
2nd term: $3^2 - 1^2$
3rd term: $4^2 - 2^2$

(*a*) Write down a similar expression for the 4th term. 1

(*b*) Hence or otherwise find the nth term in its simplest form. 3

	KU	RE

9. (*a*) Emma puts £30 worth of petrol into the empty fuel tank of her car.

Petrol costs 75 pence per litre.
Her car uses 5 litres of petrol per hour, when she drives at a particular constant speed.

At this constant speed, how many litres of petrol will remain in the car after 3 hours? **2** (KU)

(*b*) The next week, Emma puts £20 worth of petrol into the empty fuel tank of her car.

Petrol costs c pence per litre.
Her car uses k litres of petrol per hour, when she drives at another constant speed.

Find a formula for R, the amount of petrol remaining in the car after t hours. **3** (RE)

10. A badge is made from a circle of radius 5 centimetres.

Segments are taken off the top and the bottom of the circle as shown.

The straight edges are parallel.

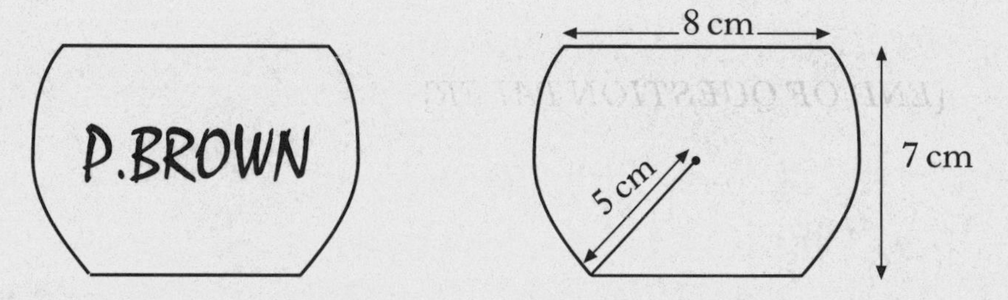

The badge measures 7 centimetres from the top to the bottom.
The top is 8 centimetres wide.

Calculate the width of the base. **5** (RE)

[Turn over

	KU	RE

11. $f(x) = 4\sqrt{x} + \sqrt{2}$

(*a*) Find the value of $f(72)$ as a surd in its simplest form. — KU 3

(*b*) Find the value of t, given that $f(t) = 3\sqrt{2}$. — RE 3

12. The height of a triangle is $(2x - 5)$ centimetres and the base is $2x$ centimetres.

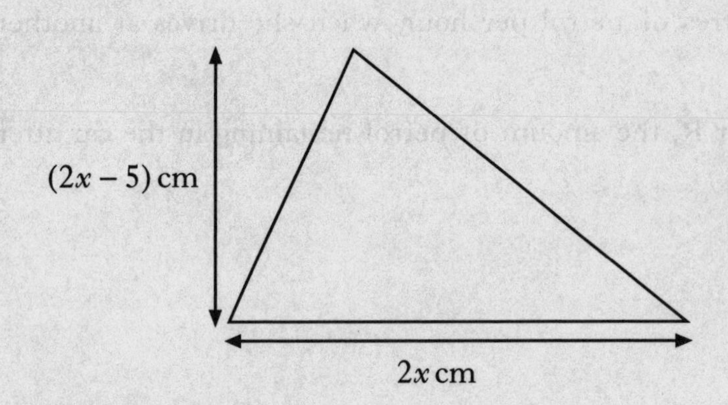

The area of the triangle is 7 square centimetres.

Calculate the value of x. — RE 5

[*END OF QUESTION PAPER*]

[BLANK PAGE]

[BLANK PAGE]

C

2500/406

NATIONAL
QUALIFICATIONS
2005

FRIDAY, 6 MAY
2.45 PM – 4.05 PM

MATHEMATICS
STANDARD GRADE
Credit Level
Paper 2

1 **You may use a calculator**.

2 Answer as many questions as you can.

3 Full credit will be given only where the solution contains appropriate working.

4 Square-ruled paper is provided.

LIB 2500/406 6/32970

FORMULAE LIST

The roots of $ax^2 + bx + c = 0$ are $x = \dfrac{-b \pm \sqrt{(b^2 - 4ac)}}{2a}$

Sine rule: $\dfrac{a}{\sin A} = \dfrac{b}{\sin B} = \dfrac{c}{\sin C}$

Cosine rule: $a^2 = b^2 + c^2 - 2bc \cos A$ or $\cos A = \dfrac{b^2 + c^2 - a^2}{2bc}$

Area of a triangle: Area $= \frac{1}{2}ab \sin C$

Standard deviation: $s = \sqrt{\dfrac{\sum (x - \bar{x})^2}{n-1}} = \sqrt{\dfrac{\sum x^2 - (\sum x)^2 / n}{n-1}}$, where n is the sample size.

KU	RE

1. $$E = mc^2.$$

Find the value of E when $m = 3{\cdot}6 \times 10^{-2}$ and $c = 3 \times 10^8$.

Give your answer **in scientific notation**. **3**

2. The running times in minutes, of 6 television programmes are:

77 91 84 71 79 75.

Calculate the mean and standard deviation of these times. **4**

3.

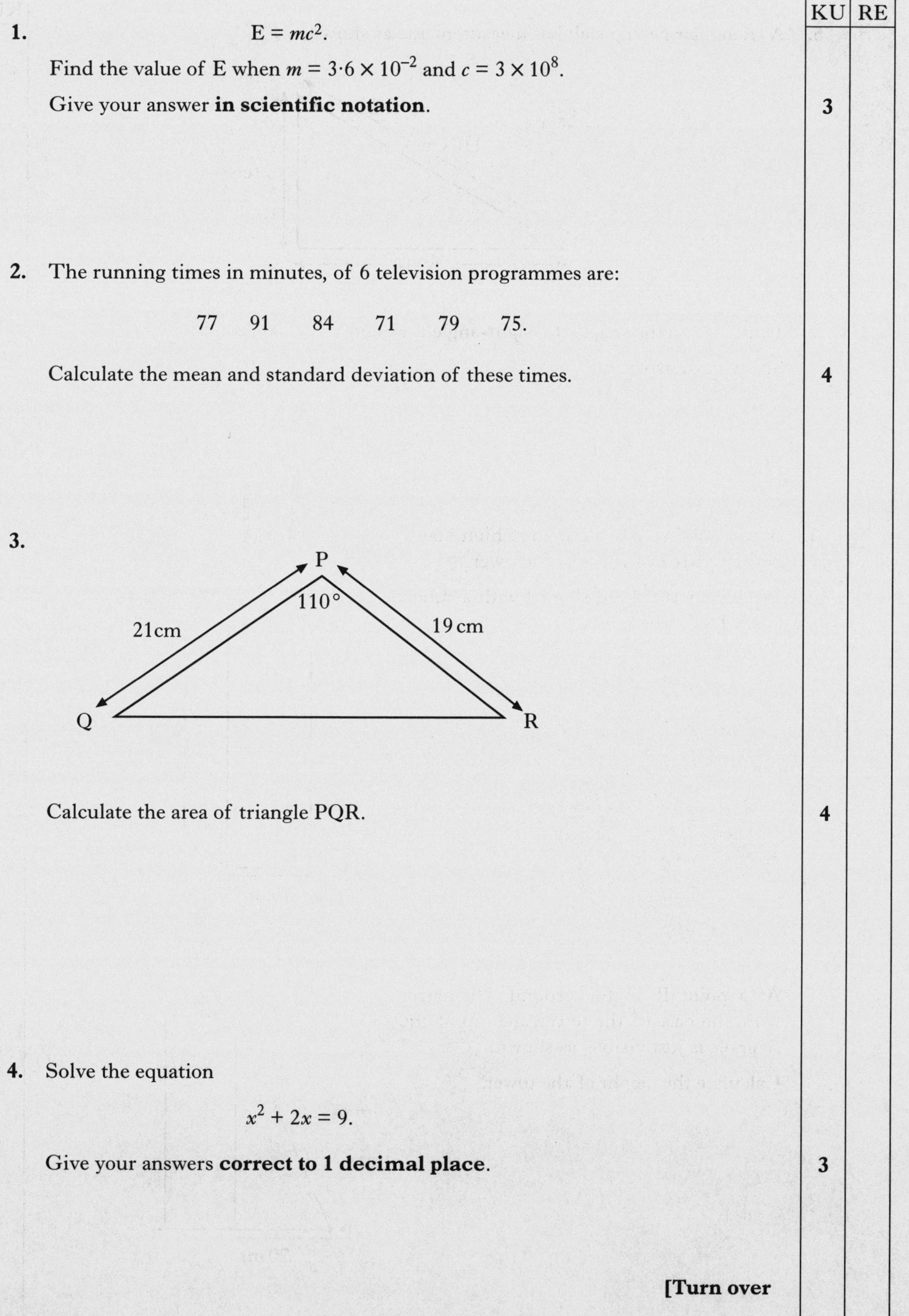

Calculate the area of triangle PQR. **4**

4. Solve the equation

$$x^2 + 2x = 9.$$

Give your answers **correct to 1 decimal place**. **3**

[Turn over

	KU	RE

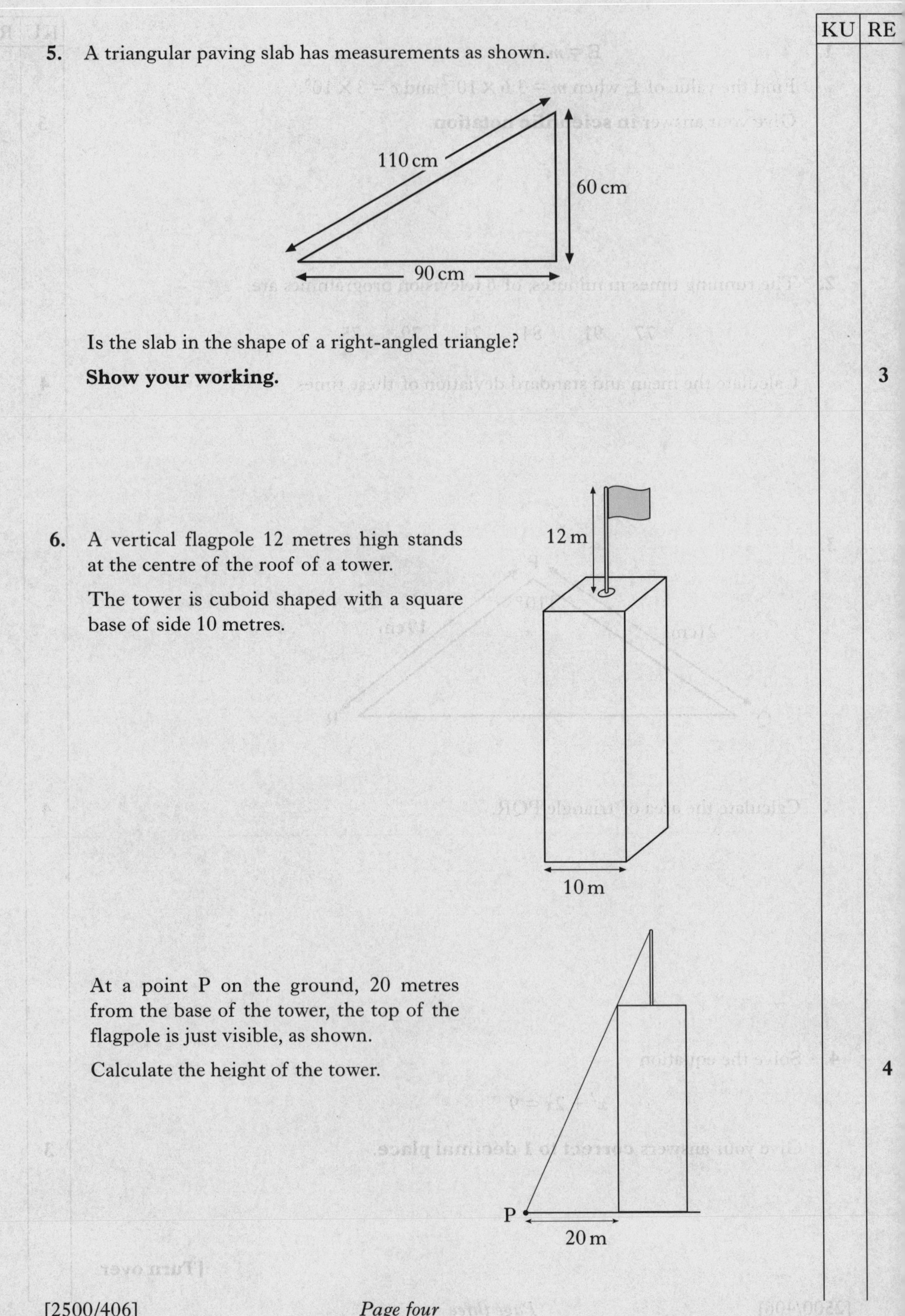

5. A triangular paving slab has measurements as shown.

Is the slab in the shape of a right-angled triangle?

Show your working. — RE 3

6. A vertical flagpole 12 metres high stands at the centre of the roof of a tower.

The tower is cuboid shaped with a square base of side 10 metres.

At a point P on the ground, 20 metres from the base of the tower, the top of the flagpole is just visible, as shown.

Calculate the height of the tower. — RE 4

	KU	RE

7. David walks on a bearing of 050° from hostel A to a viewpoint V, 5 kilometres away.

Hostel B is due east of hostel A.

Susie walks on a bearing of 294° from hostel B to the same viewpoint.

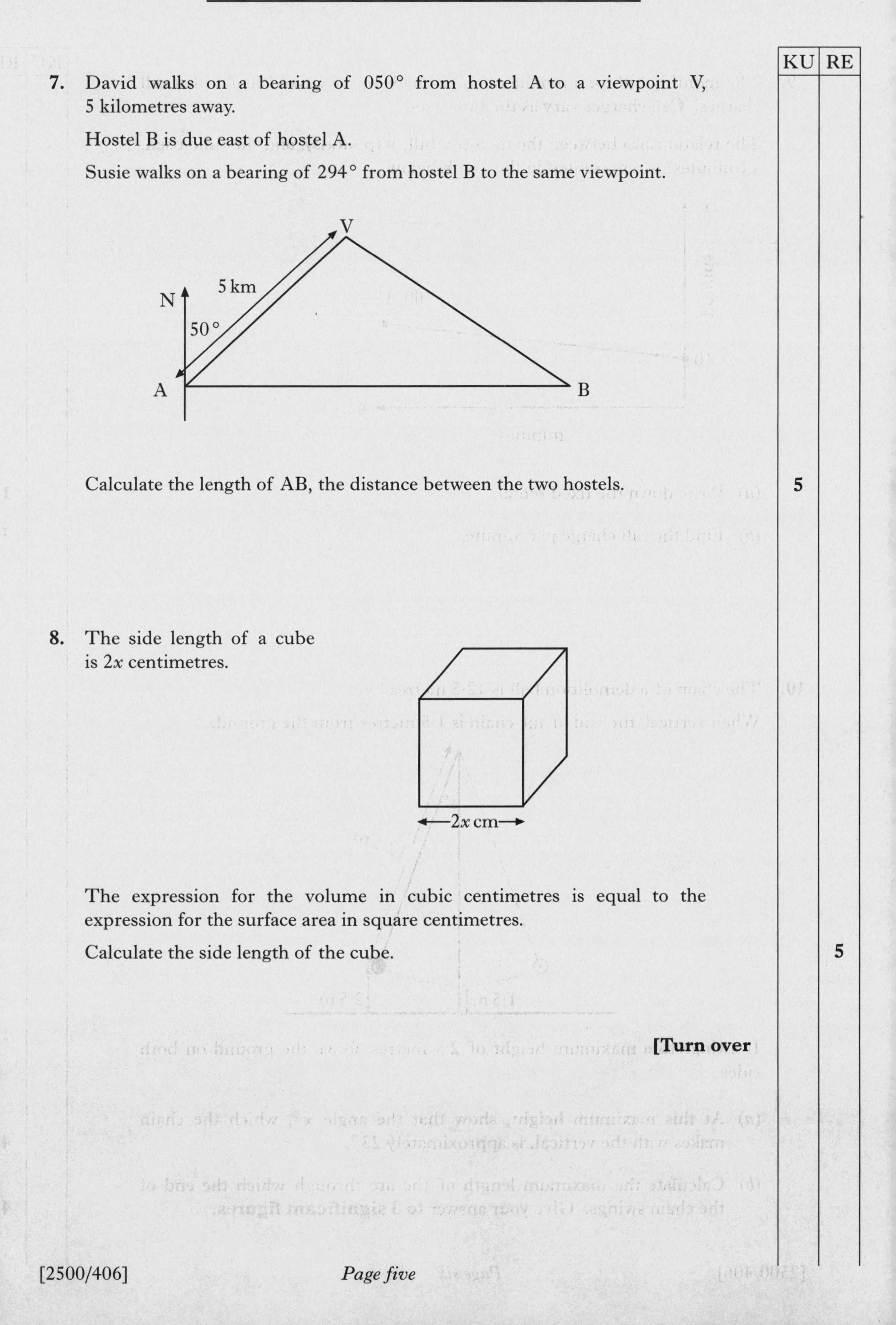

Calculate the length of AB, the distance between the two hostels. **5** (KU)

8. The side length of a cube is $2x$ centimetres.

The expression for the volume in cubic centimetres is equal to the expression for the surface area in square centimetres.

Calculate the side length of the cube. **5** (RE)

[Turn over

	KU	RE

9. The monthly bill for a mobile phone is made up of a fixed rental plus call charges. Call charges vary as the time used.

The relationship between the monthly bill, y (pounds), and the time used, x (minutes) is represented in the graph below.

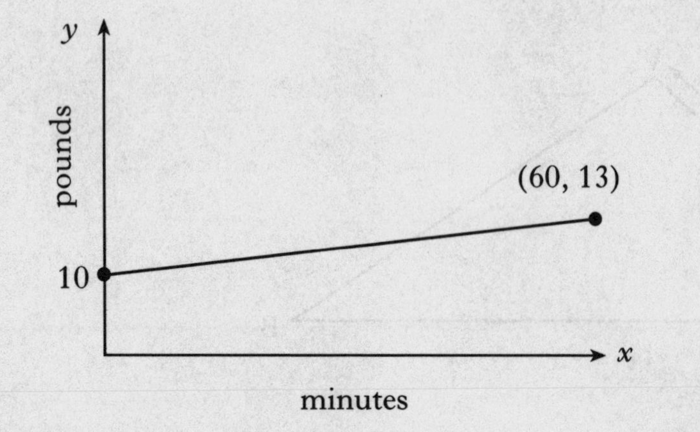

(a) Write down the fixed rental. **1**

(b) Find the call charge per minute. **3**

10. The chain of a demolition ball is 12·5 metres long.

When vertical, the end of the chain is 1·5 metres from the ground.

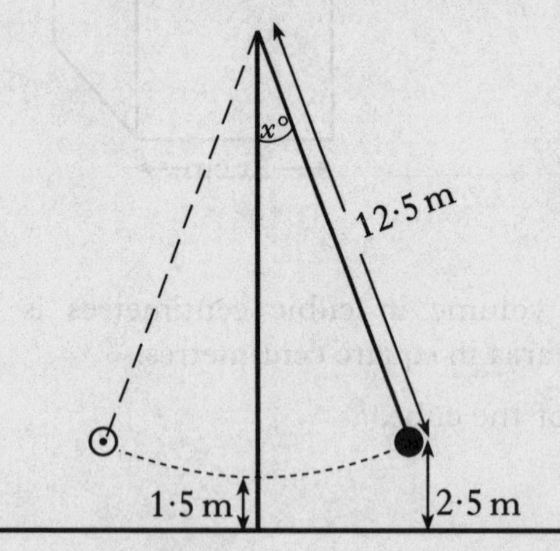

It swings to a maximum height of 2·5 metres above the ground on both sides.

(a) At this maximum height, show that the angle $x°$, which the chain makes with the vertical, is approximately 23°. **4**

(b) Calculate the maximum length of the arc through which the end of the chain swings. Give your answer **to 3 significant figures**. **4**

	KU	RE
11. (*a*) Solve algebraically the equation $\sqrt{3}\sin x° - 1 = 0 \qquad 0 \leq x < 360.$	**3**	
(*b*) Hence write down the solution of the equation $\sqrt{3}\sin 2x° - 1 = 0 \qquad 0 \leq x < 90.$		**1**

[*END OF QUESTION PAPER*]

[BLANK PAGE]

[BLANK PAGE]

[BLANK PAGE]